Let's Play
SCIENCE

Mary Stetten Carson
Illustrations by Susan Nethery

STERLING

New York / London
www.sterlingpublishing.com/kids

STERLING and the distinctive Sterling logo are registered trademarks of
Sterling Publishing Co., Inc.

Library of Congress Cataloging-in-Publication Data

Carson, Mary Stetten.
Let's play science / Mary Stetten Carson. — Rev. ed.
p. cm.
ISBN-13: 978-1-4027-3627-8
ISBN-10: 1-4027-3627-4
1. Science—Experiments. 2. Science—Study and teaching (Elementary)
—Activity programs. I. Title.

Q182.3.C368 2007
507.8—dc22

2006029602

2 4 6 8 9 7 5 3 1

Published by Sterling Publishing Co., Inc.
387 Park Avenue South, New York, NY 10016
© 2007 by Mary Stetten Carson
Illustrations © 2007 by Susan Nethery
Revised edition of *Let's Play Science*, copyright © 1979 by Mary Stetten
Distributed in Canada by Sterling Publishing
C/o Canadian Manda Group, 165 Dufferin Street
Toronto, Ontario, Canada M6K 3H6
Distributed in the United Kingdom by GMC Distribution Services
Castle Place, 166 High Street, Lewes, East Sussex, England BN7 1XU
Distributed in Australia by Capricorn Link (Australia) Pty. Ltd.
P.O. Box 704, Windsor, NSW 2756, Australia

Design by Josh Simons, SimonSays Design!

Sterling ISBN-13: 978-1-4027-3627-8
ISBN-10: 1-4027-3627-4

For information about custom editions, special sales, premium and
corporate purchases, please contact Sterling Special Sales
Department at 800-805-5489 or specialsales@sterlingpub.com.

A WORD TO THE PARENT OR TEACHER

The wide variety of activities in this book were specially chosen and designed for the enjoyment of young children under the watchful eye and timely help of a responsible parent, teacher, or guardian.

CONTENTS

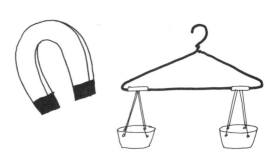
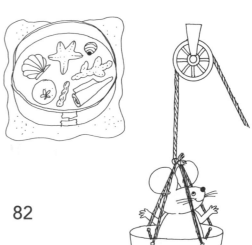

SOME EASY-TO-CARE-FOR PETS 85

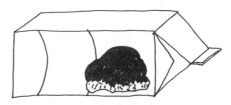

THE QUESTION IS . . .
WHAT DO WE DO FIRST?

Growing THINGS

Watch
A SEED
Start to Grow

SEEDS

You will need:

✓ dry beans such as lima or kidney beans, seeds from fruits and vegetables, or those collected outdoors (Beans are actually the seeds of certain types of plants.)

✓ paper towels

✓ a clear plastic cup

✓ water

Fill the cup loosely with crumpled paper towels.

Place several seeds into the cup between the paper and the clear side, so you can see them easily.

Drip water onto the paper towels until they are moist.

Keep the towels damp by watering every day. In a few days, the seeds should start to grow.

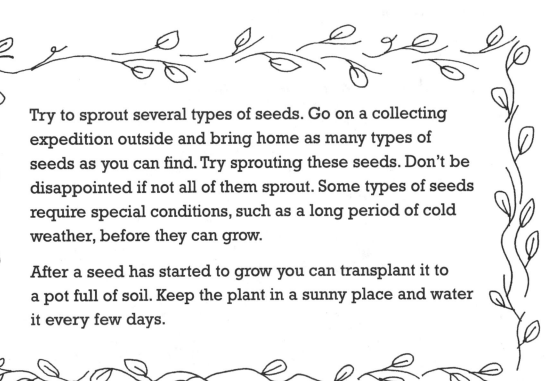

Try to sprout several types of seeds. Go on a collecting expedition outside and bring home as many types of seeds as you can find. Try sprouting these seeds. Don't be disappointed if not all of them sprout. Some types of seeds require special conditions, such as a long period of cold weather, before they can grow.

After a seed has started to grow you can transplant it to a pot full of soil. Keep the plant in a sunny place and water it every few days.

Grow (and Eat)
BEAN SPROUTS

SEEDS

SPROUT

You will need:

✓ dry beans or seeds, such as chickpeas, flax seeds, lentils, mung beans, soy beans, or alfalfa seeds (available at grocery and health food stores)

✓ a clear empty jar, or a Mason jar with a lid that has a removable center

✓ cheesecloth cut into a square large enough to fit over the top of the jar

✓ a rubber band (if using a plain jar)

✓ water

Use enough seeds to form a thin layer on the bottom of the jar. Rinse the seeds and soak them overnight in warm water.

WATER

Fasten cheesecloth over the top of the jar. (If you are using a Mason jar, remove the center section of the lid and use the outer ring to hold the cloth in place.) Pour off the water through the cheesecloth.

CHEESECLOTH

Add fresh cool water, rinse the seeds, and pour the water out. Lay the jar on its side in a dark place. Rinse the seeds two or three times a day. The sprouts will be ready to eat in three or four days. To green up the sprouts, place the jar in the sunlight for a day.

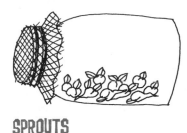

SPROUTS

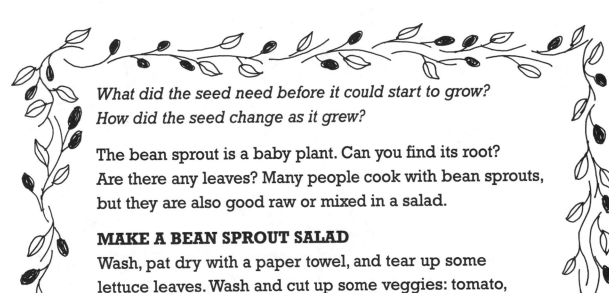

What did the seed need before it could start to grow? How did the seed change as it grew?

The bean sprout is a baby plant. Can you find its root? Are there any leaves? Many people cook with bean sprouts, but they are also good raw or mixed in a salad.

MAKE A BEAN SPROUT SALAD

Wash, pat dry with a paper towel, and tear up some lettuce leaves. Wash and cut up some veggies: tomato, carrot, onion, and whatever else you like in salad. Mix in the bean sprouts and toss with salad dressing.

Start a
SEED COLLECTION

Seeds in What You Eat

LEMON

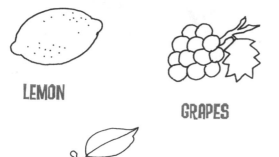

GRAPES

PLUM

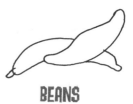

BEANS

APPLE

WATERMELON

CHERRIES

Seeds You Can Find Outdoors

MAPLE TREE SEEDS

ACORNS
(FROM OAK TREES)

MIMOSA SEED POD

CHESTNUT

DANDELION

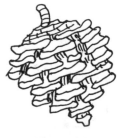

PINECONE

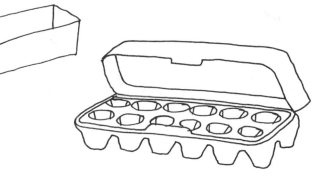

You will need:

✓ a bag for collecting seeds

✓ empty egg cartons for sorting small seeds.

✓ a box or two for larger seeds

✓ some tools for opening seeds (nutcracker, hammer, scissors)

Go on a collecting expedition in your neighborhood. Bring along a bag and collect as many types of seeds as you can find. Look for seeds from the fruits and vegetables that you eat at home.

Open some of the seeds you've found. Hard seeds, such as nuts, can be opened with a hammer or nutcracker. Other seeds can be opened with scissors, a small knife, or your fingers.

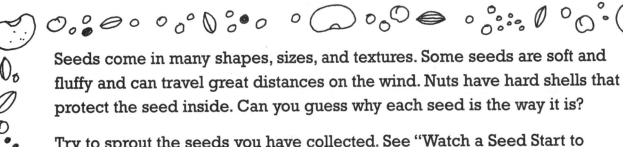

Seeds come in many shapes, sizes, and textures. Some seeds are soft and fluffy and can travel great distances on the wind. Nuts have hard shells that protect the seed inside. Can you guess why each seed is the way it is?

Try to sprout the seeds you have collected. See "Watch a Seed Start to Grow" on page 10. Use the seeds in art projects by pasting them on cardboard, or make up a seed-sorting game. Find out the names of the seeds you collected by looking for their pictures in books about trees.

Grow Plants from the Kitchen
CARROT

You will need:

✓ a fresh carrot with a green top

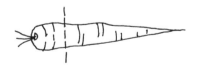

✓ a dish, pebbles, a knife

Cut off the top one or two inches of the carrot.

Put a thin layer of pebbles into the dish, place the carrot on top and add enough water just to cover the base of the carrot.

Keep in indirect sunlight and add water as needed.

POTATO

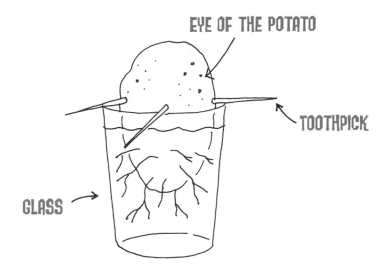

EYE OF THE POTATO

TOOTHPICK

GLASS

You will need:

✓ a potato (white or sweet) with a lot of eyes

✓ a clear container partially filled with water

✓ toothpicks

Poke toothpicks into the potato and place it in the container, with the narrow end down.

Keep white potatoes in a cool, dark place.

Keep sweet potatoes in the sunshine.

Add water as needed.

Transplant to soil after one or two months.

ONION

You will need:

✓ an onion (Try to pick one that has some roots growing at one end.)

✓ 4 toothpicks

✓ a clear container

Poke toothpicks into the onion and place it in the container with the root end down. Add water just to cover the roots.

INDIAN CORN

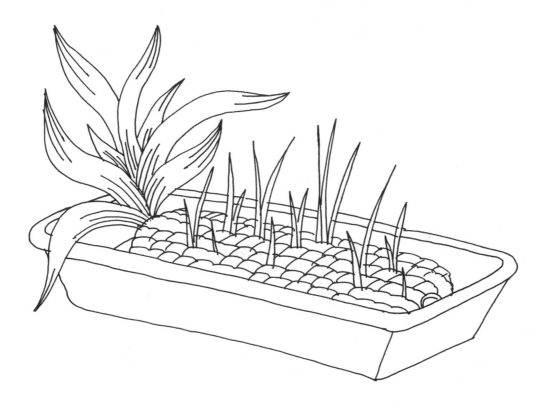

You will need:

✓ an ear of Indian corn

✓ a long shallow dish or an ice-cube tray

✓ water

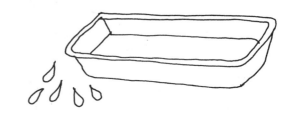

Lay the corn in the dish, and add about an inch of water. Change the water every other day.

Two Types of Leaf Collecting
PRESSED LEAVES

Collect many types of leaves. For beautiful colors, collect in the fall.

Lay the leaves between sheets of paper. To press a lot of leaves, alternate leaves and paper. Construction or blotter paper works best.

Lay a heavy stack of books on the paper, and press for one week.

Paste the leaves on heavy paper.

LEAF PRINTS

Lay leaves on newspaper with the vein side up. Paint the upper side of each leaf with tempera paint that has been thinned with water.

Transfer the leaves carefully to a clean sheet of paper, keeping the painted side up. Lay another sheet of paper on top of the leaves, and press with your hand. Lift up the top sheet of paper, and carefully peel off the leaves.

Use Your BODY

Play a TOUCH GAME

SANDPAPER

FELT

CORRUGATED CARDBOARD

CRINKLED ALUMINUM FOIL

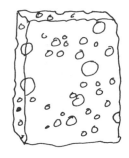

SPONGE

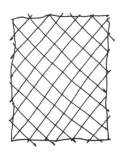

CLOTH

You will need:

- ✓ cardboard
- ✓ a blindfold
- ✓ scraps of sandpaper, sponge, cloth, etc.
- ✓ white glue or glue stick
- ✓ scissors

Cut the cardboard into rectangles about the size of playing cards. Glue a different-textured material to one side of each card.

Put on a blindfold, and have a friend select a card for you to feel. How does the material feel? *Soft? Bumpy? Rough? Smooth?* Return the card to its place.

Remove the blindfold, and try to guess which card you felt.

Your skin is covered with tiny nerve endings that send messages to your brain. Your brain can then tell you how a thing feels.

Try touching the cards with different parts of your body. Can you feel just as well with your nose? Your toes? Your elbow?

How many words can you think of to describe how different things feel? Hot, cold, wet, dry, prickly, smooth, spongy—see how many more you can name.

Mystery TOUCH BOX

You will need:

✓ a cardboard box

✓ scissors

✓ some old socks

✓ felt-tip pens or paint (to decorate the box)

✓ small objects with distinct shapes

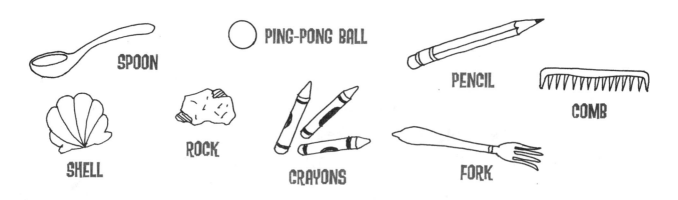

SPOON

PING-PONG BALL

PENCIL

COMB

SHELL

ROCK

CRAYONS

FORK

Cut an opening in the box large enough for a hand to fit through. Decorate the box with markers, paints, or stickers.

Put a different object in each sock, and tie the sock closed.

Put all the socks into the box.

Take a sock from the box. Feel it. What do you think is inside? Open the sock and find out if you guessed correctly. Ask friends over to play the game.

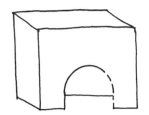

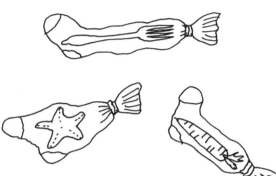

I was right.

How did you guess what was in your sock? Try to describe how it felt. Was it soft, hard, smooth, bumpy? What was its shape?

Play a TASTE GAME

POPCORN

CHOCOLATE CHIPS

MINI-MARSHMALLOWS

CRACKERS

RAISINS

CARROT SLICES

Look at your tongue in the mirror. If you look very carefully, you will see tiny bumps on the top side. These bumps are called *taste buds*, and they are what you use for tasting.

What are some of your favorite foods?
Can you describe how they taste?

You will need:

✓ many different types of food, cut into bite-size pieces

CHEESE

COOKIE

ORANGE

GRAPES

APPLE

CEREAL

✓ cups, one for each type of food

✓ a blindfold

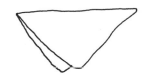

Put on the blindfold, and have a friend place one type of food in your mouth.

How does it taste?
How does it feel? *Is it . . .*

SWEET?   SOUR?   CRUNCHY?   CHEWY?   SPICY?

Can you guess what it is?

Play a SMELL GAME

Your nose can smell many things. Have you ever smelled a fire burning? Or gasoline at a gas station? Does your mother wear perfume? Sometimes you can even tell what is cooking for dinner by its smell.

Don't try to play the Smell Game if you have a bad cold. When your nose is stuffy, it is often very hard to tell one smell from another.

You will need:

✓ empty plastic containers with lids

✓ foods with distinct smells

CHOCOLATE CHIPS

LEMON

CHEESE

ONION

APPLE

VANILLA

PEANUT BUTTER

Place a small amount of food in each container and put on the lids.

Allow them to stand for a while at room temperature so that the smells can grow strong.

Open one container a crack and sniff. Do not peek at the food inside. Can you guess what the food is by its smell? Look inside and find out if you were right.

Play a MUSIC GAME

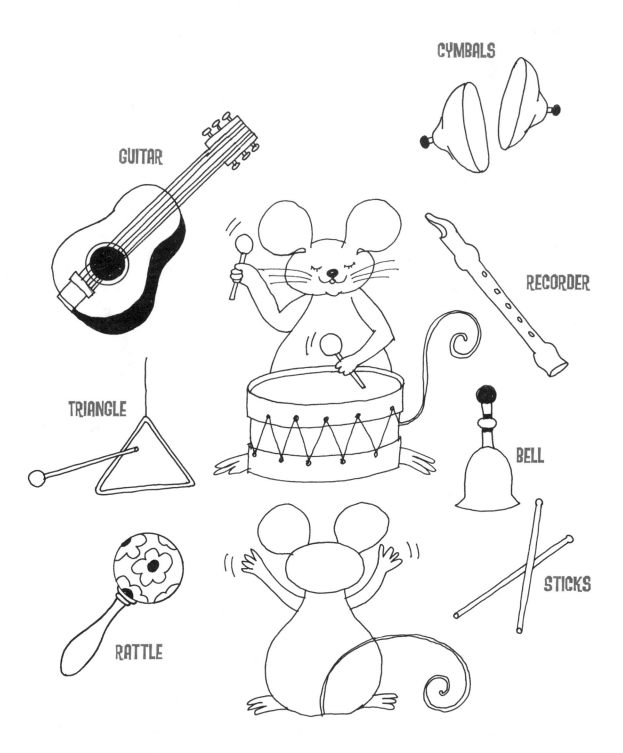

CYMBALS

GUITAR

RECORDER

TRIANGLE

BELL

STICKS

RATTLE

You will need:

✓ musical instruments, such as a bell, rattle, guitar, and drum

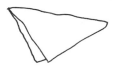

✓ a blindfold

Put on a blindfold. Have a friend play one of the instruments. How did it sound? Can you guess which instrument you heard?

What words can you use to describe how the instrument sounded? Did it bang or crash or ring or tap or tinkle?

How many ways can you find to play the instrument? Can you play it loud? Soft? Fast? Slow?

When you listen, you are using your ears. Can you hear just as well if you have your fingers in your ears?

Sit very still for a minute or two. How many noises can you hear? Did a car honk its horn? Did a dog bark? Are people talking to each other in the next room?

Try playing the Music Game with instruments you have made yourself. Create your own instruments, or follow instructions on the pages that follow.

Make a RATTLE

You will need:

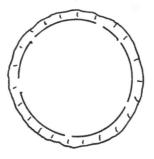

✓ dry beans or pebbles

✓ a paper plate

✓ a stapler with staples

✓ crayons or felt-tip pens

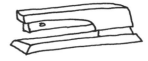

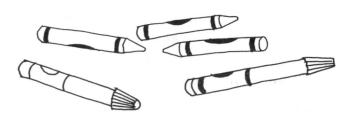

Fold the paper plate in half.

Drop in some beans or pebbles.

Staple the plate together near the edge.

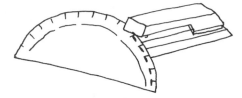

Play a RATTLE GAME

Make several sets of rattles, filling each set of two with the same material.

FILLED WITH BEANS

FILLED WITH PENNIES

FILLED WITH SMALL PEBBLES

FILLED WITH PAPER CLIPS

On one side of each rattle, draw a picture of what is inside. Mix the rattles together with the picture sides down.

Find the matching pairs by shaking the rattles and comparing the sounds. Check by looking at the pictures.

Make a KAZOO

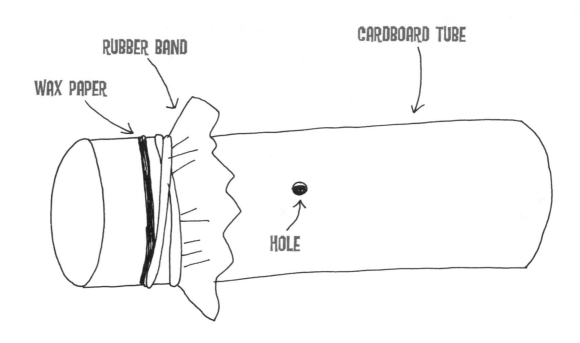

WAX PAPER

RUBBER BAND

CARDBOARD TUBE

HOLE

Why does the kazoo make such a funny sound?

Place your finger gently on the wax paper while you play the kazoo. What do you feel? The wax paper tickles your finger, because it is moving back and forth very quickly.

All sounds are caused by something vibrating. When you hit a drum, the top of the drum vibrates. When you pluck a guitar string, the string vibrates.

Make several kazoos so that you and your friends can play them together. Form a kazoo band and play all your favorite songs.

You will need:

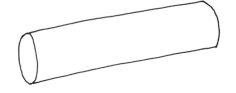

✓ a cardboard tube from a roll of toilet paper or paper towels

✓ some wax paper to cover one end of the tube

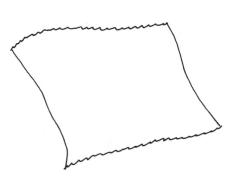

✓ a rubber band

✓ a pencil

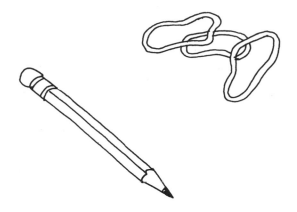

Cover one end of the tube with wax paper, and secure it with a rubber band.

Use a pencil to poke a hole in the tube near the wax paper and allow air to escape.

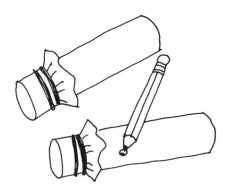

To play the kazoo, put the open end to your mouth and "sing" your favorite song, using "du-du-du" for the words.

Make a STRINGED THING

RUBBER BAND

You will need:

✓ scissors

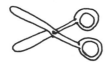

✓ rubber bands

✓ an empty milk carton

Wash out the milk carton and open up the top.
Cut down the sides to form four flaps.

Cut out one side of the milk carton, leaving
a small rim on all four sides.

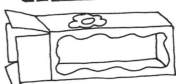

Fold the open end of the carton so that it lies flat.

Stretch several rubber bands around the carton...
and play it!

Try using different sizes of rubber bands on the Stringed
Thing. Do the strings make different sounds? Can you tell
which type of rubber band produces a higher sound and
which a lower sound?

What other types of instruments have strings? A violin?
A guitar? Does a piano have strings?

Make a Hose Telephone

Hello, do you hear me? Can you guess how the hose telephone works?

You may think that the hose is empty, but it isn't—it's filled with air. When you talk into one end, the air starts to vibrate. That means it moves back and forth very quickly. The vibrating air carries your voice along through the hose until it comes out the other end—into your friend's ear.

This type of phone was actually used a long time ago in large houses, so people could talk to each other from different rooms. It's still a common type of telephone on ships.

Try using the phone to talk to a friend in another room. Use the phone with several friends, and try to guess whom you are talking to. But remember, never yell through a hose telephone—the loud noise might hurt your friend's ear.

You will need:

✓ a hose (10, 20, even 30 feet long!) Buy hosing at a hardware store, or use an old garden hose with the ends cut off.

✓ 2 plastic containers open at one end

✓ a small knife

✓ heavy tape

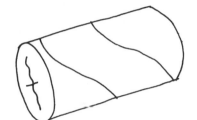

Cut a cross in the base of the container.

Push one end of the hose through the slit and out the open end of the container.

Wrap tape several times around the hose near the end.

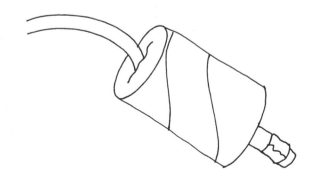

Pull the hose back into the container.

Repeat the above with the other end of the hose.

Make a RAINBOW

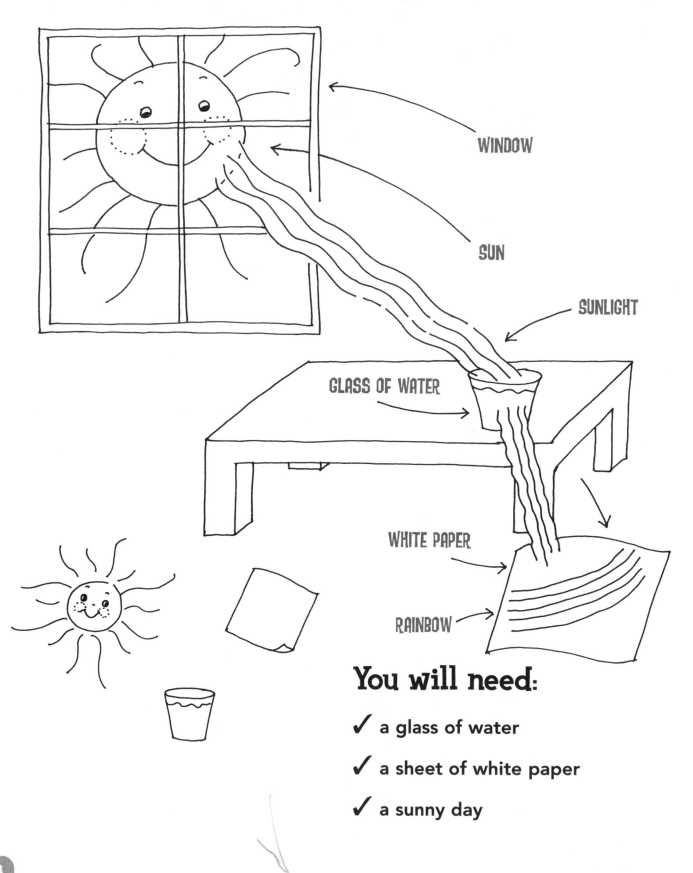

WINDOW

SUN

SUNLIGHT

GLASS OF WATER

WHITE PAPER

RAINBOW

You will need:

✓ a glass of water

✓ a sheet of white paper

✓ a sunny day

Fill the glass almost to the top with water. Place the glass so that it is half on and half off the edge of a table and so that the sun shines directly through the water and onto a sheet of white paper on the floor.

Adjust the paper and the glass until a rainbow forms on the paper.

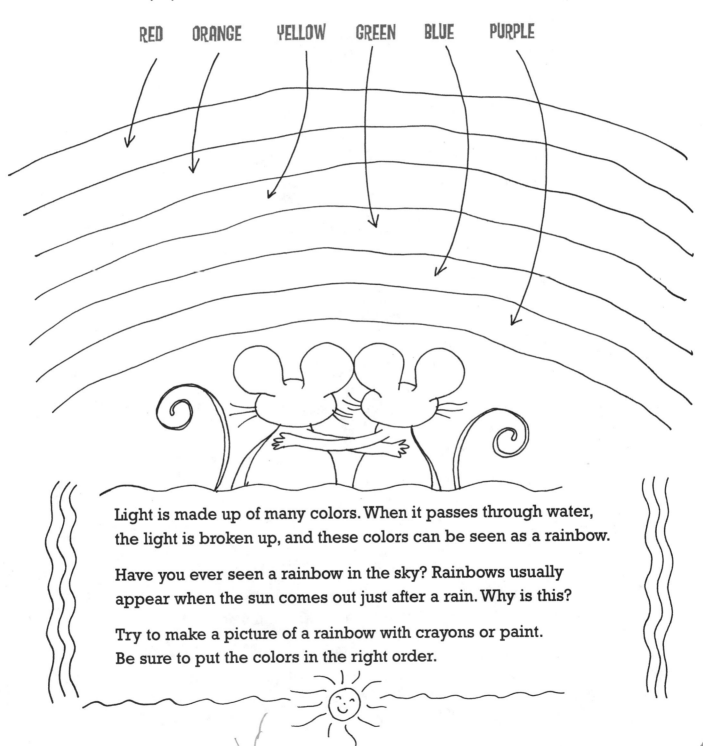

RED ORANGE YELLOW GREEN BLUE PURPLE

Light is made up of many colors. When it passes through water, the light is broken up, and these colors can be seen as a rainbow.

Have you ever seen a rainbow in the sky? Rainbows usually appear when the sun comes out just after a rain. Why is this?

Try to make a picture of a rainbow with crayons or paint. Be sure to put the colors in the right order.

Play a Shadow
GUESSING GAME

Can you explain what causes shadows? Could you make a shadow if it was completely dark? Turn on a single lamp in your room, and try making shadows on the walls and ceiling.

Look for shadows when you play outside. Is it easier to see shadows on a sunny or a cloudy day?

You will need:

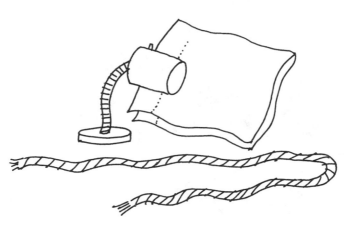

✓ **a sheet**

✓ **a lamp**

✓ **a rope**

✓ **objects with different shapes**

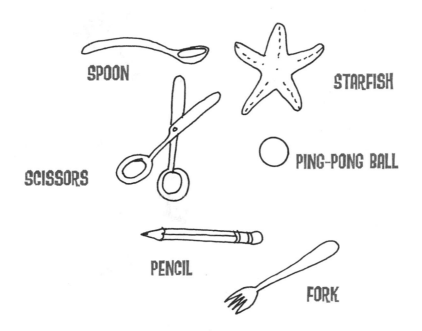

SPOON

STARFISH

SCISSORS

PING-PONG BALL

PENCIL

FORK

String the rope across the room, and hang the sheet over it. Have your friends sit on one side of the sheet, and shine the light on the other side. Darken the room.

Stand between the light and the sheet, and hold up one of the objects so that the shadow shows on the other side of the sheet. Have your friends try to guess what the object is by the shape of its shadow.

Move the object toward the light. What happens to the size of the shadow it casts on the sheet?

Make a
SHADOW THEATER

You will need:

✓ a rope

✓ a sheet

✓ a lamp

✓ heavy paper or cardboard

✓ pipe cleaners or long sticks

✓ masking tape

✓ scissors

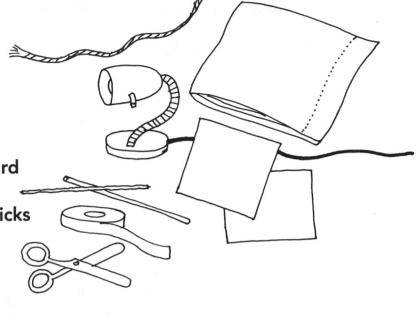

Cut figures of animals, people, trees, etc., from heavy paper or cardboard.

Tape a bent pipe cleaner or a stick to the back of each figure.

Set up the sheet and light as you did for the Shadow Guessing Game. Darken the room.

TAPE

PIPE CLEANER

Move the puppets behind the sheet. Make the shadows big or small by moving the puppets back and forth between the sheet and the lamp.

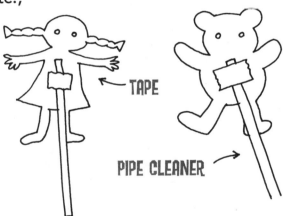

Make a
SHADOW PICTURE

You will need:

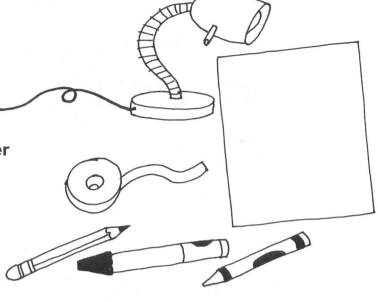

✓ a lamp

✓ a large sheet of white paper

✓ masking tape

✓ a pencil or felt-tip pen

Have your friend stand near a wall, and shine a light so that the shadow of the face's profile appears on the wall.

Tape a sheet of paper on the wall, and adjust the light and your friend so that the shadow falls onto the paper.

Carefully trace the outline of the shadow with a pencil or felt-tip pen. Your friend must stand very still for you to draw a good portrait.

Remove the paper from the wall, and write your friend's name on the back.

Before the camera was invented, shadow pictures were a very popular way of making portraits.

Draw shadow pictures of your friends, making sure to write the subject's name on the back of each picture. Mix up the pictures and let your friends try to guess the identity of each one.

Make a ROCKET (with a Balloon)

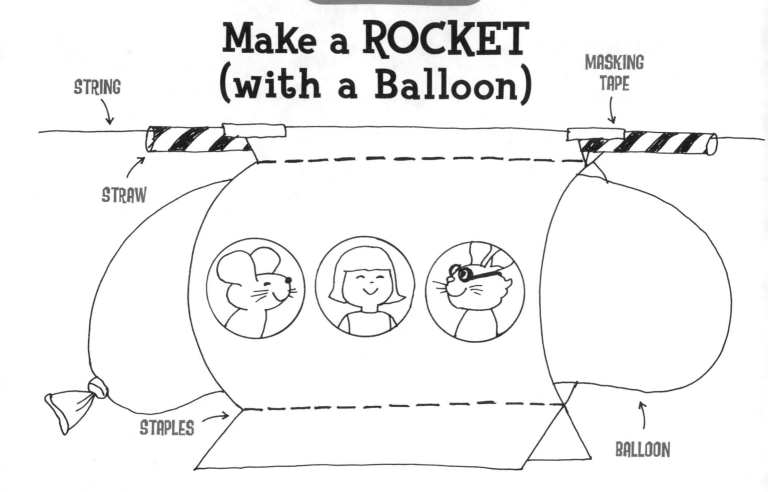

STRING

MASKING TAPE

STRAW

STAPLES

BALLOON

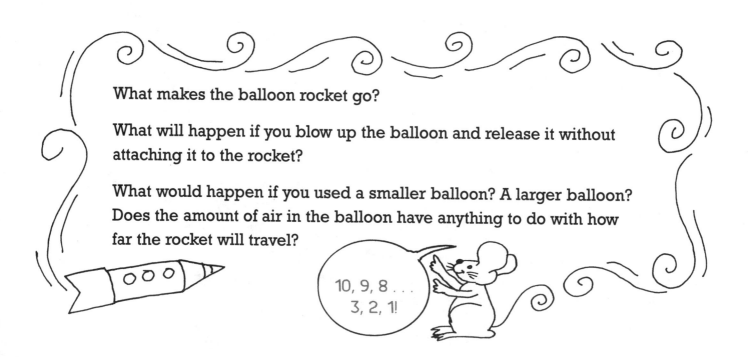

What makes the balloon rocket go?

What will happen if you blow up the balloon and release it without attaching it to the rocket?

What would happen if you used a smaller balloon? A larger balloon? Does the amount of air in the balloon have anything to do with how far the rocket will travel?

10, 9, 8 . . . 3, 2, 1!

You will need:

✓ a piece of string—long enough to reach across the room

✓ several long balloons

✓ a drinking straw

✓ a piece of construction paper

✓ felt-tip pens or crayons

✓ a stapler with staples

✓ masking tape

Thread the string through the straw, and tie the string so that it stretches tightly across the room.

Fold the construction paper in half, and decorate it to resemble a rocket ship.

Hang the folded paper over the straw and tape it in place.

Staple the paper together just under the straw and again near the bottom.

Tape a deflated balloon to the inside of the paper so that the opening of the balloon sticks out from the back end of the rocket. Blow up the balloon and release to fly the rocket.

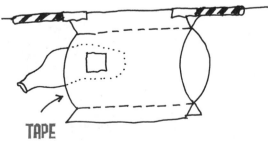

TAPE

Make a PINWHEEL

You will need:

✓ a square piece of paper

✓ scissors

✓ a pin

✓ a stapler with staples

✓ masking tape

✓ a straw

Fold the paper into a triangle.

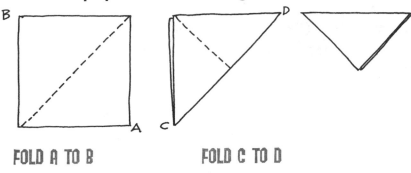

FOLD A TO B   FOLD C TO D

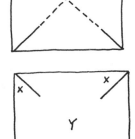

Unfold it, and cut along the fold lines only halfway to the center.

Bend the corners (x) of the paper into the center (y) to make air pockets. Do not crease.

Staple the pinwheel together near the center.

Stick a pin through the center of the pinwheel (front to back) and into the straw. Wind masking tape around the top part of the straw. Make sure to cover the sharp point of the pin completely. Cover the top of the straw and both sides of the pin to hold the paper away from the straw.

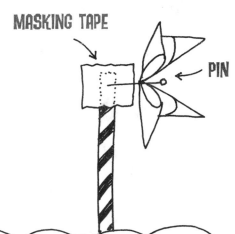

MASKING TAPE

PIN

What makes the pinwheel move?

You cannot see air, but when you blow, the air moves and pushes the pinwheel round and round. Moving air can also push things like sailboats, windmills, and balloon rockets. See "Make a Rocket (with a Balloon)."

Can you make the pinwheel go around without blowing on it? One way is to move the pinwheel through the air by swinging it around or running with it. You can also hold the pinwheel in front of a fan or near an open car window as you ride along.

Make a Paper
AIRPLANE

You will need:

✓ a small sheet of paper

✓ a stapler with staples

✓ scissors

Try to throw a piece of paper that has not been folded. How does it fly? Can you explain why the folded airplane flies better than the plain sheet of paper?

There are many ways of folding paper into airplanes. Fold and test your own models.

Fold a piece of paper in half lengthwise. Open up the paper and lay it on the table with the crease down the middle.

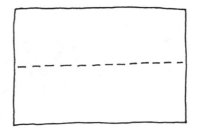

Fold two corners until they meet at the center crease.

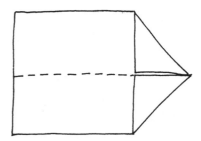

Fold the same two corners again to form a sharper point, then fold the plane on the center crease.

FOLD LINE

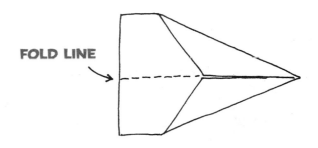

Staple through the layers near the bottom, then fold each side down to the bottom edge to form the wings.

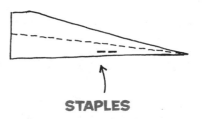

STAPLES

Lift the wings to each side, and cut off the tip of the plane.

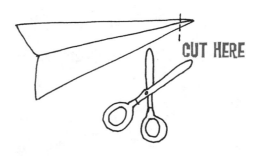

CUT HERE

Make a
HELICOPTER

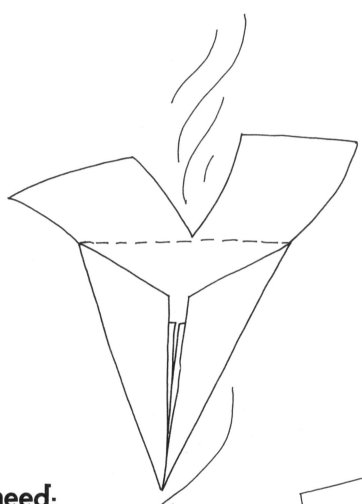

You will need:

✓ a small piece of paper
   (3 by 5 or 2 by 3 inches)

✓ scissors

✓ crayons or felt-tip pens

Crease the paper down the center lengthwise, then unfold.

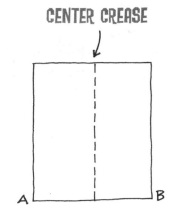

CENTER CREASE

Fold points A and B until they meet on the center crease.

Fold points C and D until they meet on the center crease.

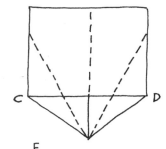

Cut from point E to F down the center crease.

Fold one flap slightly forward and one flap slightly backward.

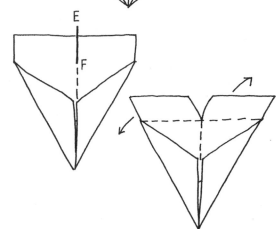

Drop the helicopter from a high place.

What would happen if you dropped the paper at each stage of folding? Try it. How does it fall before you start to fold? How does it fall with a point but no flaps? Does the angle of the flaps affect how the finished helicopter falls?

Try using different sizes of paper. Try throwing it like an airplane. What happens?

Decorate the helicopter if you wish.

Make a PARACHUTE

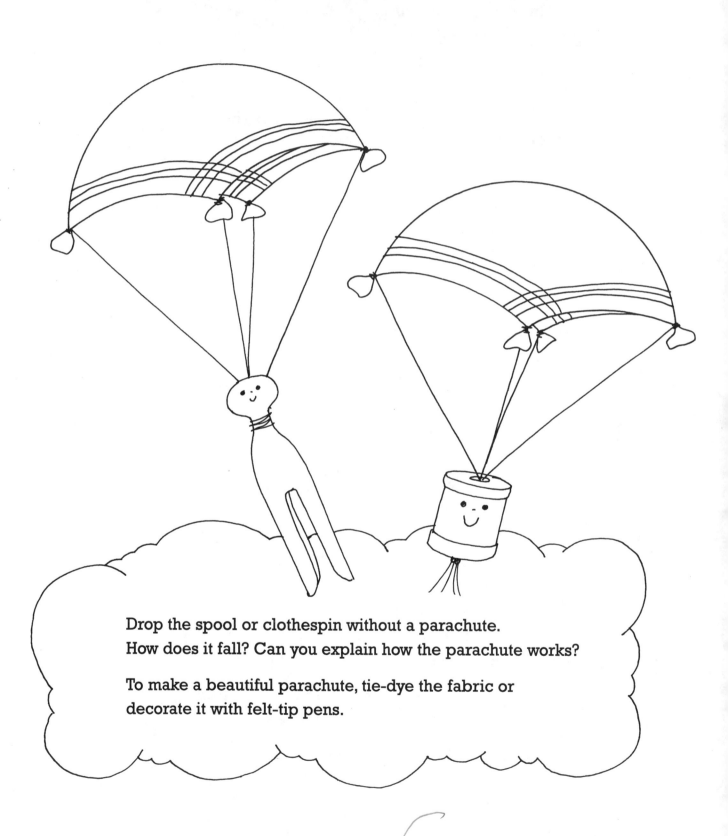

Drop the spool or clothespin without a parachute.
How does it fall? Can you explain how the parachute works?

To make a beautiful parachute, tie-dye the fabric or
decorate it with felt-tip pens.

You will need:

✓ a square piece of cloth
   (a handkerchief size works well)

✓ string

✓ scissors

✓ a clothespin, spool, or small doll

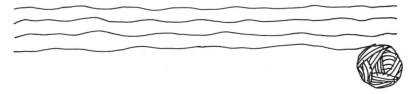

Cut four pieces of string of equal length.

Tie one piece of string to each corner of the cloth.

If you're using a spool, thread all four strings through the hole and tie them together. Add a bead, if the knot pulls through.

If you're using a clothespin, tie the four strings together and then tie on the clothespin with a separate piece of string.

Draw on faces.

Blow BUBBLES

You will need:

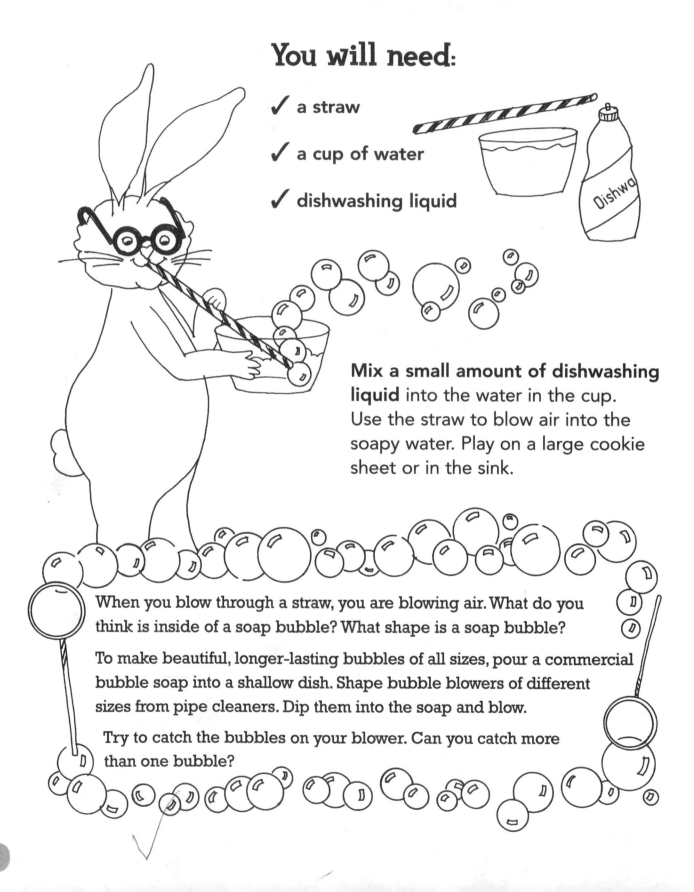

✓ a straw

✓ a cup of water

✓ dishwashing liquid

Mix a small amount of dishwashing liquid into the water in the cup. Use the straw to blow air into the soapy water. Play on a large cookie sheet or in the sink.

When you blow through a straw, you are blowing air. What do you think is inside of a soap bubble? What shape is a soap bubble?

To make beautiful, longer-lasting bubbles of all sizes, pour a commercial bubble soap into a shallow dish. Shape bubble blowers of different sizes from pipe cleaners. Dip them into the soap and blow.

Try to catch the bubbles on your blower. Can you catch more than one bubble?

Make a BLOW PICTURE

You will need:

✓ **food coloring or tempera paint thinned with water**

✓ **a straw**

✓ **white paper or a paper plate**

Put a large drop of paint on the paper. Blow at it through the straw. Continue blowing at wet spots and turning the paper to create a picture. Add several colors.

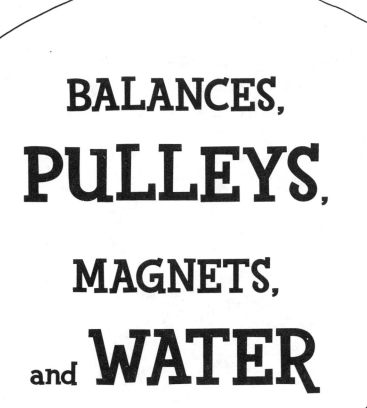

Make a
COAT-HANGER BALANCE

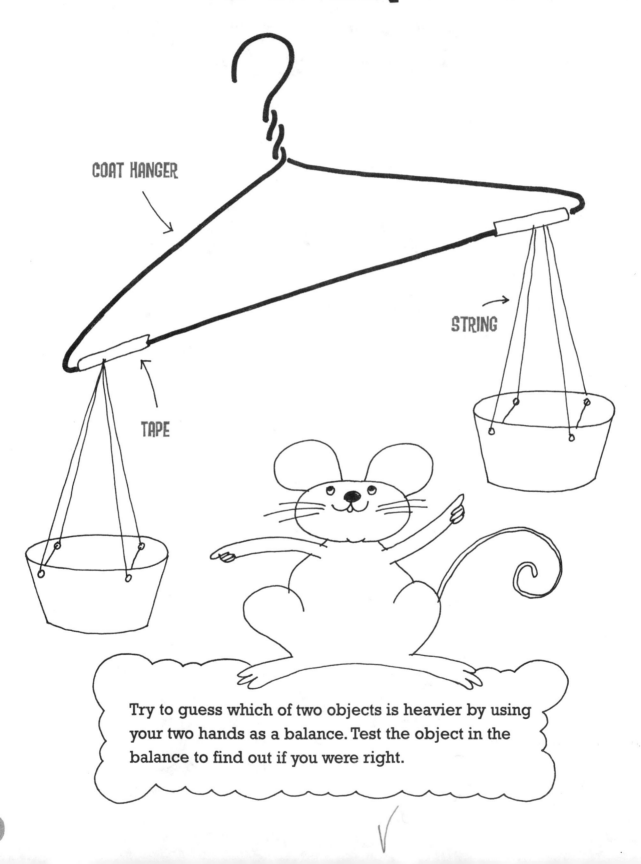

COAT HANGER

STRING

TAPE

Try to guess which of two objects is heavier by using your two hands as a balance. Test the object in the balance to find out if you were right.

You will need:

✓ a coat hanger

✓ string

✓ 2 identical plastic containers

✓ a nail

✓ scissors

✓ masking tape

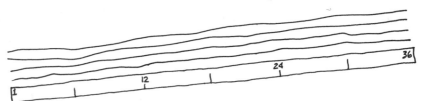

Cut four lengths of string, each a yard long.

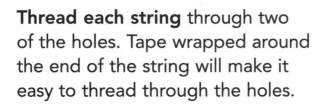

Poke four holes in each container with the nail. The holes should be evenly spaced near the rim.

Thread each string through two of the holes. Tape wrapped around the end of the string will make it easy to thread through the holes.

Tie the four strings together near the ends.

Tie a container to each end of the hanger, and secure to the hanger with masking tape.

Make a BALL BALANCE

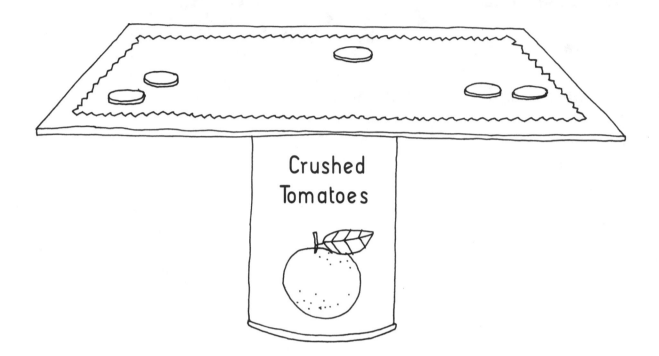

You will need:

- ✓ an empty can
- ✓ newspaper
- ✓ a rubber ball, small enough to fit into the can
- ✓ heavy cardboard
- ✓ scissors
- ✓ felt
- ✓ glue
- ✓ pennies

Fill the can most of the way up with crumpled newspaper, and set a ball on the newspaper.

Cut the cardboard into a square, and cut two pieces of felt the same size.

Glue the felt to both sides of the cardboard.

Balance the board on the ball.
Use pennies or other coins as weights, placing them on the board so that the board tilts or remains level.

Play a balance game: Take turns placing coins on the balance. The player who makes the board fall is the loser.
How many coins can you balance?

Make balance boards in other shapes.

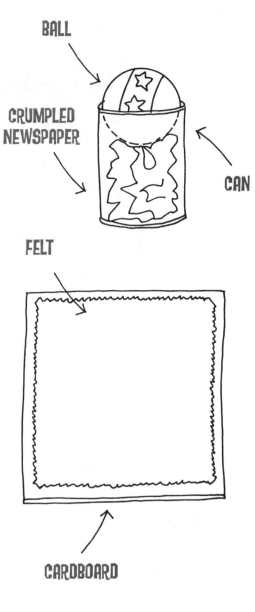

BALL

CRUMPLED NEWSPAPER

CAN

FELT

CARDBOARD

COINS

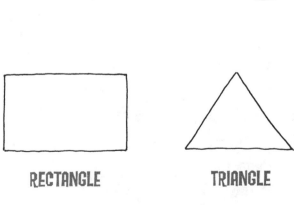

CIRCLE

RECTANGLE

TRIANGLE

Put PULLEYS to work

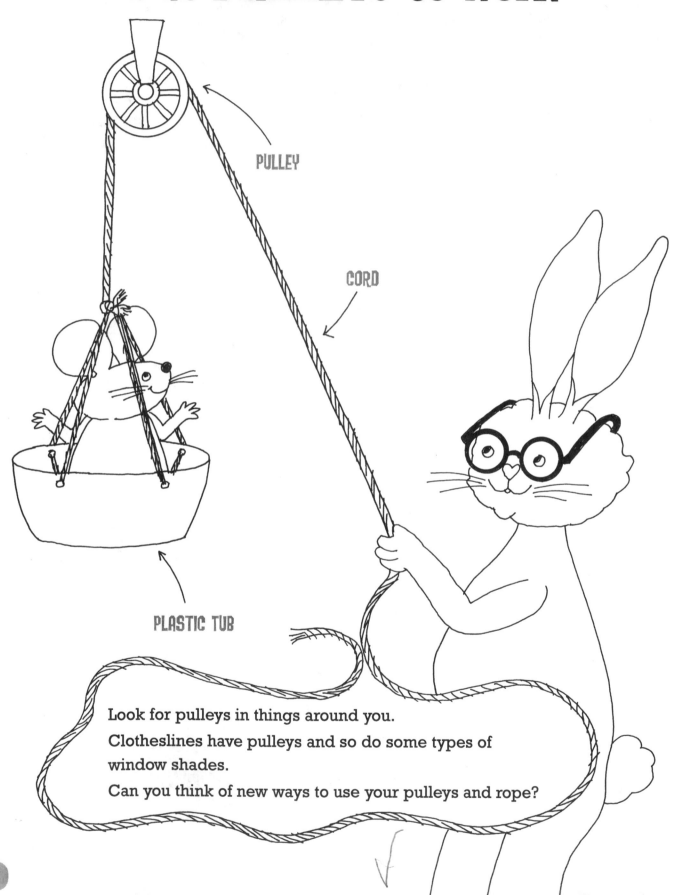

PULLEY

CORD

PLASTIC TUB

Look for pulleys in things around you.

Clotheslines have pulleys and so do some types of window shades.

Can you think of new ways to use your pulleys and rope?

You will need:

✓ pulleys (available at hardware stores)

✓ a small plastic container

✓ string or cord

✓ scissors

✓ a nail

Use the nail to poke four holes in the container.
The holes should be near the rim and evenly spaced.
Thread two strings through the holes (as in "Make a
Coat-Hanger Balance").

Tie the short strings to one long string, and thread the
long string through the pulley. Tie the pulley to a high
place. Put a doll in the elevator and give it a ride.

How many ways can you find to use a rope and pulleys?

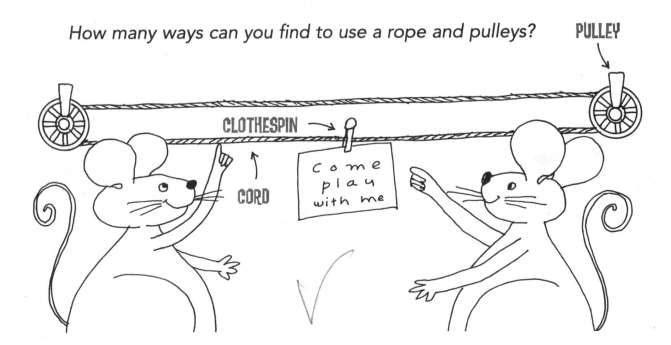

PULLEY

CLOTHESPIN

CORD

come
play
with me

Play a Magnet
SORTING GAME

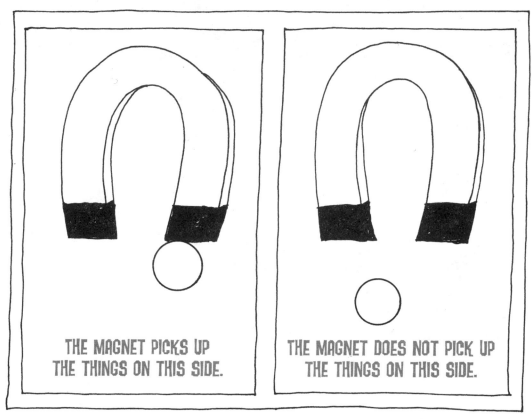

THE MAGNET PICKS UP
THE THINGS ON THIS SIDE.

THE MAGNET DOES NOT PICK UP
THE THINGS ON THIS SIDE.

PLAYING BOARD

MAGNET

You will need:

✓ a sheet of white paper

✓ felt-tip pens or crayons

✓ a magnet

✓ a number of small objects made of various materials

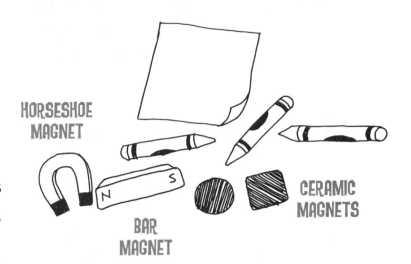

HORSESHOE MAGNET

BAR MAGNET

CERAMIC MAGNETS

Draw a playing board on the paper.

Select an object and guess whether or not the magnet will pick it up. Test the object and put it on the proper side of the playing board.

Repeat until all the objects are sorted.

Look at the objects that you have sorted. Can you explain why the magnet picked up certain things but not others?

Keep a box of magnets and small metal objects such as screws and paper clips. By playing with your magnets, you'll discover more about what they can do.

WARNING Careful! Don't store magnets, or play with them, near computers, credit cards, or other electronic devices. Magnetic fields may cause damage.

Play a
MAGNET FISHING
Game

You will need:

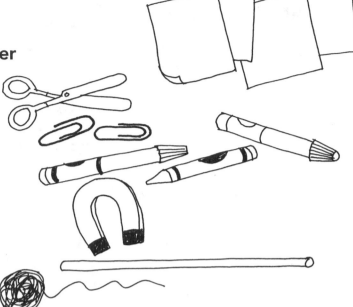

✓ colored construction paper

✓ scissors

✓ paper clips

✓ felt-tip pens or crayons

✓ a magnet

✓ a stick or dowel

✓ string

Cut fish from different-colored paper, and draw on eyes and mouths with felt-tip pens.

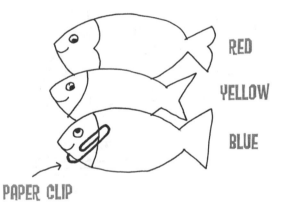

RED

YELLOW

BLUE

Attach a paper clip near the mouth of each fish.

PAPER CLIP

Tie one end of the string to the dowel and the other to the magnet to make a fishing pole.

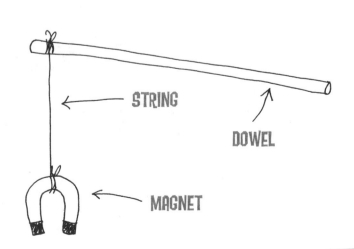

STRING

DOWEL

MAGNET

Scatter the fish on the floor and go fishing!

Science in the BATHTUB

Body Wash

Soap

You will need:

✓ empty plastic bottles, paper cups, a funnel, an eggbeater, and anything else you can find around your home that would be fun to play with in the tub (WARNING Never play with anything breakable in the tub!)

✓ a nail

✓ food coloring

✓ bubble-bath soap

Use the nail to poke holes in some of the plastic bottles and paper cups.

Add small amounts of food coloring to the water in some of the bottles, and use the colored water in your play.

Put some bubble-bath soap in the tub water, and use an eggbeater to beat up a big mound of bubbles.

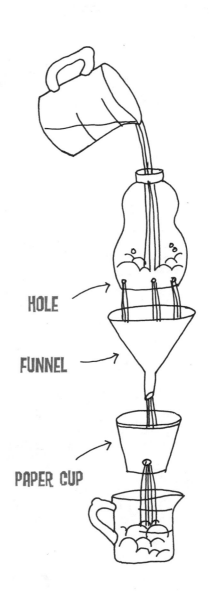

HOLE

FUNNEL

PAPER CUP

Play a SINK-FLOAT Game

EVERYTHING THAT SINKS
GOES ON THIS SIDE.

EVERYTHING THAT FLOATS
GOES ON THIS SIDE.

Sink

Float

PLAYING BOARD

BASIN OF WATER

THINGS TO TEST

You will need:

✓ white cardboard

✓ felt-tip pens

✓ transparent contact paper

✓ a tub or large pot of water

✓ assorted objects that sink or float

PENCIL BUTTON ROCK CRAYON PINECONE CORK SCISSORS SCREW

Draw a sink-float playing board on the cardboard, and cover the board with clear contact paper.

Guess whether an object will sink or float. Test by putting the object in the water. Then put it on the correct side of the board.

Repeat until all the objects are sorted according to whether they sink or float.

Look over the objects you have sorted. Can you explain why some of them sink and others float?

Float a toy boat. Drop small weights, such as pebbles or coins, one by one into the boat. How many can you drop in before the boat sinks?

MAKE AN ICY Treat

You will need:

✓ orange juice, grape juice, or apple juice

✓ an ice-cube tray

✓ 12 craft sticks

✓ aluminum foil

✓ a freezer

Pour some juice in an ice cube tray. Don't fill it all the way, because the liquid will expand when it freezes.

JUICE

ICE-CUBE TRAY

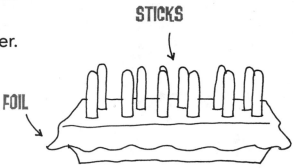

Cover the tray with a double layer of aluminum foil. Crimp it all around the edges to keep the foil in place. Push a craft stick through the foil into the center of each cube section. Carefully place the tray in the freezer.

STICKS

FOIL

When the ice treats have frozen, peel off the foil and carefully remove as many treats as you need. Leave the remaining ones in the tray, and return them to the freezer for another day.

Make a TAKE-HOME
Beach Sculpture

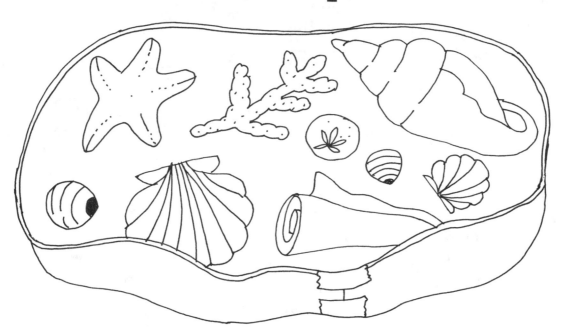

You will need:

✓ a small bag of plaster of Paris

✓ a strip of cardboard (about 5 by 40 inches)

✓ masking tape

✓ a cup for measuring

✓ a bucket

✓ a spoon or stick for mixing

✓ a beach or sandbox

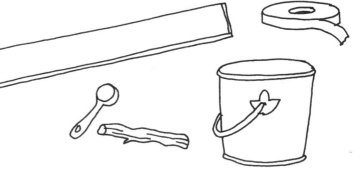

You can make a beach sculpture at the beach or at home in the sandbox.

Bend the strip of cardboard and overlap the ends. Tape them together.

Shape the cardboard into a circle, oval, square, or rectangle. Press the cardboard frame into the sand. Arrange beach shells, stones, driftwood, sea glass, etc. on the sand inside the frame. Place them "best" side down. (Collect other treasures for a sandbox sculpture.)

In the bucket, mix one cup of water for every two cups of plaster of Paris. If you are at the beach, water from the ocean will work fine.

WATER   PLASTER   PLASTER

Mix the plaster of Paris with a spoon or stick until it is smooth, then pour it into the mold. The plaster of Paris should be at least 1½ inches thick. Mix extra, if you need more.

Let the plaster of Paris dry and harden. This will take about half an hour. Turn the sculpture over and remove the cardboard mold.

Some
Easy-to-Care-For
PETS

GERBILS

You will need:

✓ a glass aquarium with a screen cover

✓ a water bottle

✓ a shallow dish

✓ bedding (aspen shavings or paper pulp)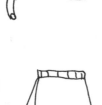

✓ an exercise wheel

✓ gerbil food

✓ a cardboard tube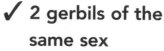

✓ 2 gerbils of the same sex

Spread a 1-inch layer of bedding over the floor of the aquarium. Add the water bottle, food dish, exercise wheel, and gerbils.

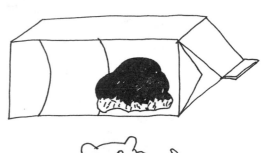

Give the gerbils tubes from rolls of paper towels or toilet paper, or empty cardboard boxes with openings cut in them. The gerbils will chew up the cardboard for nesting material.

Gerbils are very clean animals, so you will not have to clean the aquarium more than once every few months. Allow the cedar chips and chewed cardboard to build up until there is enough for the gerbils to dig burrows.

Gerbils enjoy company, so it's a good idea to buy two—of the same sex, unless you want to start raising gerbils.

In addition to eating regular gerbil food, gerbils love special treats such as peanuts, potato chips, or small pieces of carrot and lettuce. They also enjoy getting out of the aquarium now and then for some exercise. A good safe place to let them loose is in the bathroom with the door shut. (Hang a "Careful—Gerbil Run" sign on the door.)

The best way to pick up a gerbil is to grasp it by the base of the tail with one hand and lift it up on the palm of the other hand. If you always handle it gently, your gerbil will become a tame and friendly pet.

CRICKETS

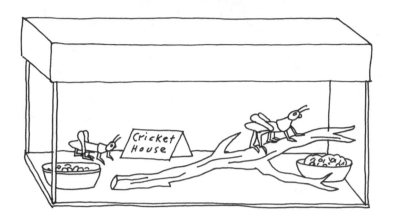

You will need:

- ✓ crickets (available at pet stores as reptile food)

- ✓ a glass aquarium or a clear plastic storage box with a lid

- ✓ 2 shallow dishes, saucers, or jar lids

- ✓ a sponge

- ✓ a branch

- ✓ a small piece of paper

- ✓ scissors

- ✓ crackers, bread crumbs, or cereal

- ✓ water

Cut the sponge so that it fits into one of the dishes or lids. Moisten the sponge with water. The crickets will drink by sipping from the sponge.

Put a very small amount of crumbled cracker, cereal, or bread in the other dish.

Put in a branch for the crickets to climb on and a small piece of folded paper for them to hide under.

Keep the aquarium or box covered so that the crickets can't escape.

SPONGE

CRICKET HOUSE

Crickets shed their skins, so don't be surprised if one day you discover some empty skins in the aquarium or box. Can you guess why this happens?

If your crickets are happy in their new home, they might start to "sing," which they do by rubbing their wings together.

ANTS

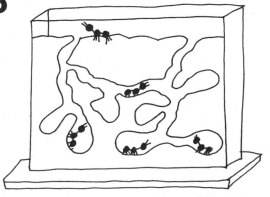

The easiest way to start a successful ant colony is to buy an ant-farm kit at a pet store. The kit comes with a coupon, which you mail in to receive your ants.

CATERPILLARS

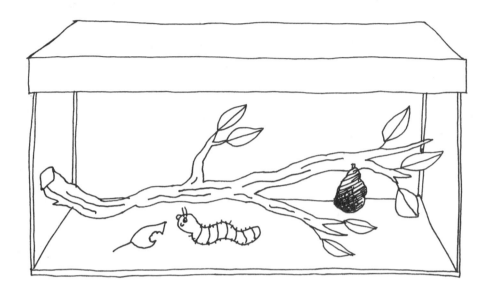

You will need:

✓ a clear plastic storage box with a lid

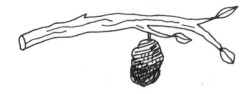

✓ a caterpillar

✓ leaves and small twigs from where you found the caterpillar

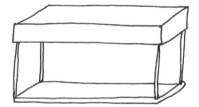

✓ or some kind of cocoon attached to a small branch

Put the caterpillar into the box with the twigs and leaves. Feed the caterpillar by providing fresh leaves every day. After a while, the caterpillar will build a protective covering around itself and change into a butterfly or a moth.

Do not disturb the cocoon. When the butterfly or moth emerges, take the box outside and remove the cover so that it can fly away.

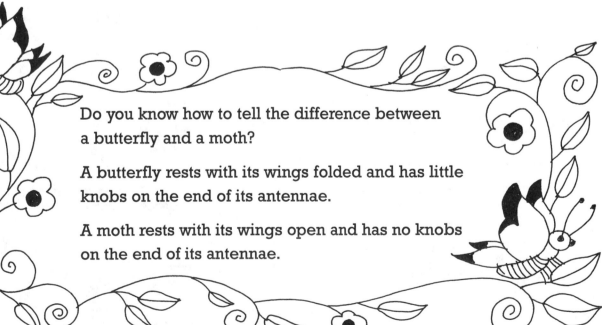

Do you know how to tell the difference between a butterfly and a moth?

A butterfly rests with its wings folded and has little knobs on the end of its antennae.

A moth rests with its wings open and has no knobs on the end of its antennae.

WORMS

You will need:

✓ a pail and shovel

✓ a jar with a lid

✓ a hammer and nail

✓ sand or gravel,
and soil

The best time to dig for worms is after a rain shower when the soil is still moist. Dig up some dirt and go through it with your fingers. You might have to dig in several spots before you find any worms.

Use the hammer and nail to make several holes in the lid of the jar. Put a layer of loose sand or gravel at the bottom of the jar. Then fill the jar loosely with dirt from the area where you found the worms.

Put the worms in the jar. Remember to keep the soil moist.

Earthworms are good for plants. As they tunnel through the earth, they loosen and enrich the soil. When you are finished keeping your worms, it would be a good idea to release them in a local garden.

INDEX

THE END

ABOUT THE AUTHOR

MARY STETTEN CARSON came from a family of scientists, but only discovered her own love of science and education over many summers as a student at the Woods Hole Children's School of Science on Cape Cod.

After receiving a B.A. from Oberlin College and a Masters in Early Childhood Education from Bank Street College of Education in New York City, Mary combined her interests by becoming a specialist in preschool science. *Let's Play Science*, her first book, grew out of her experiences with young children and was followed by *The Scientific Kid*. Mary has worked as a teacher and consultant in many schools in New York City. She teaches a graduate level workshop at Bank Street College of Education and runs the science center at the West Side YMCA Nursery School.

Mary Carson lives in Manhattan with her husband, Michael, and her gray cat, Oliver, and is the proud mother of Matthew, who works with computers, and Maggie, who plays the banjo.